Juares Elias Santos Junior

Molecular Epidemiology

Juares Elias Santos Junior

Molecular Epidemiology

Enteropathogen Giardia

ScienciaScripts

Imprint

Cover image: www.ingimage.com

This book is a translation from the original published under ISBN 978-3-330-77000-3.

Publisher:
Sciencia Scripts
is a trademark of
Dodo Books Indian Ocean Ltd. and OmniScriptum S.R.L publishing group

120 High Road, East Finchley, London, N2 9ED, United Kingdom
Str. Armeneasca 28/1, office 1, Chisinau MD-2012, Republic of Moldova, Europe
Managing Directors: Ieva Konstantinova, Victoria Ursu
info@omniscriptum.com

Printed at: see last page
ISBN: 978-620-8-62719-5

SUMMARY

Dedication

My wife Rafaela Santos

My parents, Mary Bellio and Juares Santos

"*Life doesn't deserve us to worry so much.* "

(Marie Sklodowska Curie)

1 INTRODUCTION

Giardiasis is an intestinal infection caused by a flagellated protozoan of the genus *Giardia,* located in the small intestine of mammals, birds, reptiles and amphibians. This parasite was perhaps the first protozoan to be described in human feces.[1,2] Its first description was attributed to Anton Van Leewenhoek (1681), who noticed "mobile animalcules" in his feces, but it was Lambl, in 1859, who described it in more detail. The genus was created by Kunstler (1882) when he observed a flagellate in the intestines of anuran tadpoles.[3]

According to Levine *et al.* (1980) the systematic classification of this protozoan is included in the Kingdom Protista, subkingdom Protozoa, of the phylum Sarcomastigophora, in the sub-phylum Mastigophora, class Zoomastigophorasida, order Diplomonadorida, sub-order Diplomonadina and family Hemamitidae.[4]

Using morphological criteria, three species of *Giardia* were proposed*: Giardia duodenalis* (*G. duodenalis*), which infects various mammals, including humans, birds and reptiles; *Giardia muris*, which infects rodents, birds and reptiles; and *Giardia agilis*, which infects amphibians. From studies using electron microscopy, other species have been proposed: *Giardia psittaci* and *Giardia ardeae*, described in parakeets and blue herons, respectively, and *Giardia microti*, found in rodents known as muskrats and field mice.[2]

The species *Giardia lamblia* became widely accepted in the 1970s.[5] Since the 1980s, some researchers have encouraged the use of the name *Giardia duodenalis* [6,7] and in the 1990s, the name *Giardia intestinalis* was widely used by other researchers.[2,8,9] Despite disagreement as to the validity of their taxonomy, all these species continue to be used by various authors. Therefore, *G. lamblia*, *G. duodenalis and G. intestinalis* have been used synonymously, particularly for isolates of human origin.[10,11]

This parasite is currently one of the most studied organisms, not only because of its ubiquity as a parasite, but also because of its importance in human and animal health, as it is the causative agent of diarrhea and can contribute to nutritional deficiencies and weight gain difficulties. An increase in the number of cases of infection has been observed, particularly in children who attend or live in collective environments, such as nurseries and schools[12,13].

1.1. Morphology and Structure

G. duodenalis has two evolutionary forms, the trophozoite and the cyst. The trophozoite inhabits the host's small intestine, causing giardiasis. It has a pyriform shape, is mobile and has two nodes located in the anterior region (Figure 1). It has a very simple intracellular organization, no mitochondria, a primitive vesicular secretory system which is considered to be the archetype of the Golgi secretory apparatus of higher organisms.(14) There is currently a consensus that this parasite originated as a primitive branch of the eukaryote lineage which diverged before the acquisition of mitochondria.(15)

The cytoskeleton *of G. duodenalis* is made up of organelles composed of microtubules, which are the ventral disc, the basal bodies, four pairs of flagella and a pair of median bodies.(5,13) The flagella, eight in number, originate from the basal bodies located close to the nucellus and emerge from the anterior, posterior, ventral and caudal regions of the trophozoite. The median bodies are located in the middle of the body and dorsally to the caudal flagella and are made up of a bundle of microtubules. Their function is unknown and their morphology varies according to the species of *Giardia.*(5) The dorsal face of this protozoan is smooth and convex, while the ventral face is concave, presenting a suction cup-like structure known as the ventral, adhesive or suctorial disc. The disk has been considered an important structure for the parasite's adhesion to the mucosa.(16)

The cyst (Figure 2) is the parasite's form of resistance and is infective to hosts. Two to four nuclei, basal bodies, median bodies and structural elements of the ventral disc and flagella are visible.(5,8,9) Electron microscopy has provided new information on the structures and possible function of the various organelles of the cyst and trophozoite.(8)

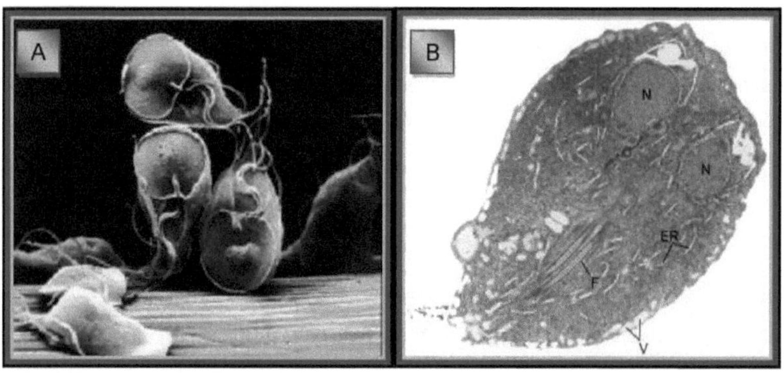

Figura 1. (A) *G. duodenalis* trophozoites by scanning electron microscopy (SEM). Source: http://www.e-parasites.co.uk/images/giardia.gif (B) Cross-section of *G. duodenalis* trophozoite visualized by transmission electron microscopy (TEM). Nucleus (N), flagellum (F), vacuoles (V), endoplasmic reticulum (ER).(5)

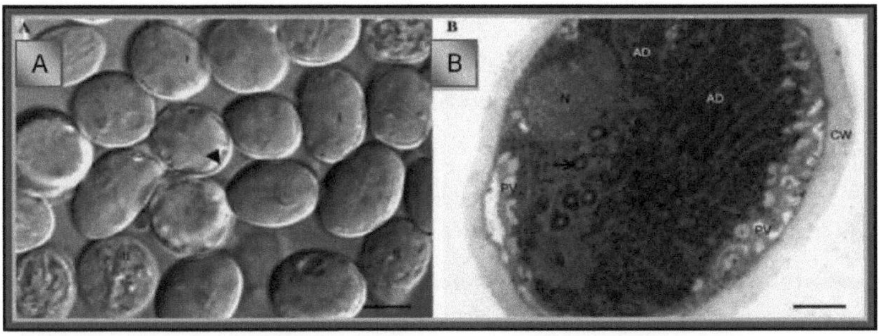

Figura 2. *G. duodenalis* cysts obtained *in vitro*. (A) Optical microscopy (OM) of the cysts showing their thick wall. (B) MET of the cyst showing the cystic wall homogeneously distributed around the cell membrane (CW). Flagellar axonemes (arrow), nucleus (N), ventral disc (AD) and peripheral vesicles (PV).(17)

1.2 Biological cycle

In the biological cycle of *G. duodenalis*, the host becomes infected by ingesting cysts of the parasite, which are present in water, contaminated food or through interpersonal contact(9) (Figure 3). After ingestion, they pass through the acidic environment of the stomach and disembark in the initial portion of the small intestine. Each cyst produces two trophozoites united by cytoplasm which immediately divides to separate them (Figure 4). Each trophozoite attaches itself to the intestinal mucosa and divides by binary division (Figure 5). The process of encystment of the trophozoites takes place in the final portion of the small intestine. The flagella shorten, the cytoplasm condenses and a cystic membrane is produced. Newly formed cysts have two nuclei, but then there is nuclear division inside the cyst and four nuclei are formed. Soon afterwards, the adhesive disk, basal bodies, median bodies and locomotor apparatus (flagella) are duplicated, resulting in two trophozoites which remain connected by the cytoplasm inside the cyst(16). These cysts are eliminated from the host's body along with their feces and can contaminate the hands of the parasitized person, food and water(9).

It is now accepted that, in medium-intensity infections, the number of cysts eliminated in the feces can reach 300 million per day.(18) Parasitized individuals do not eliminate cysts continuously; this intermittent elimination of cysts in the feces is called a "negative period" and can last an average of 10 days. In addition, the pattern of excretion of cysts varies from individual to individual; in low excretors, stool samples can remain negative for 20 consecutive days.(3)

In the environment, cysts can survive for up to two months if the temperature and humidity conditions are favorable, and it is not known whether they are all readily infectious, but there is evidence that they can take up to seven days to become mature and infectious[19] Millions or billions of cysts can be found in a single stool. It should be noted that in diarrheal stools there is a large presence of trophozoites, but in solid stools cysts predominate. It is also known that some of the trophozoites eliminated in the diarrheal material can lead to infection if they survive the action of the gastric secretion[5,8,20]

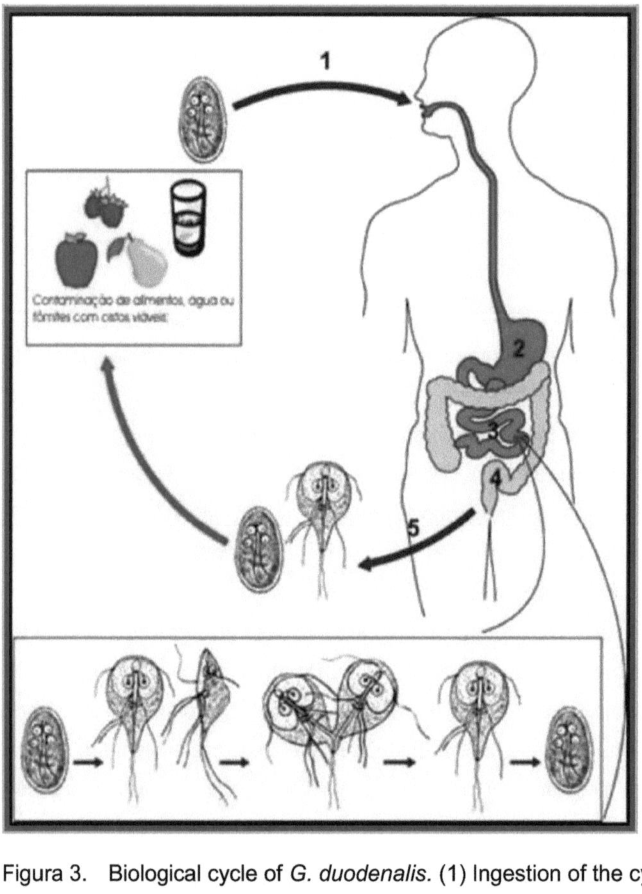

Figura 3. Biological cycle of *G. duodenalis.* (1) Ingestion of the cyst through contaminated food. (2) Cyst in the stomach area making its way to the small intestine. (3) Detachment and multiplication of the parasite. (4) Elimination through the feces. (5) Cystic form and trophozoite in the environment.

http://www.fcfrp.usp.br/dactb/Parasitologia/Arquivos/Genero Giardia files/i mage007.jpg [21]

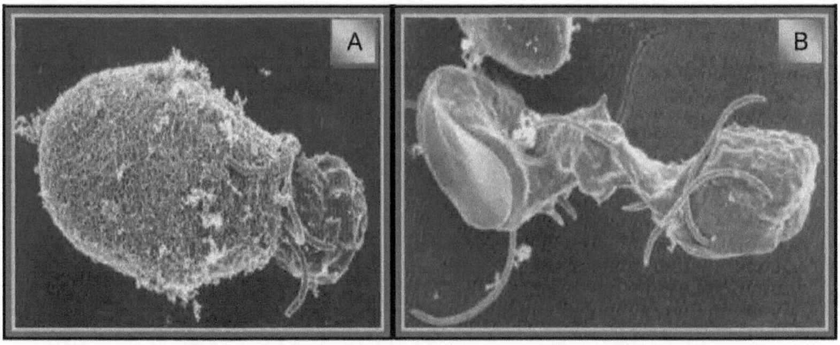

Figura 4. (A) Trophozoites united by a single cytoplasm emerging from the cyst

(B) Trophozoites after exiting the cyst and completing the cytoplasmic division process (SEM).[22]

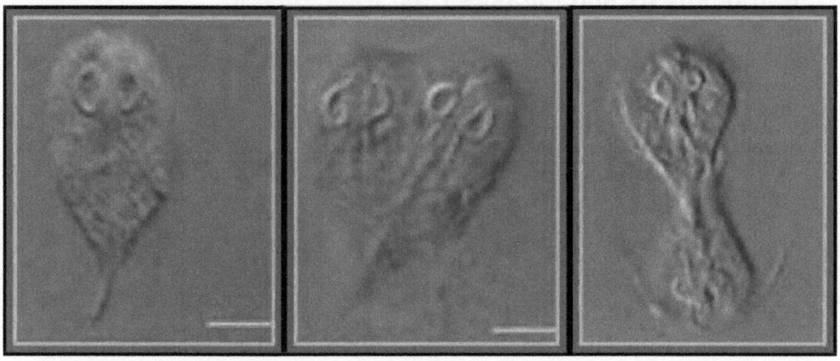

Figura 5. The process of mitosis carried out by *G. duodenalis* using microscopy with an immunofluorescence (IM) system.[23]

In experimental infections of volunteers, implantation of the parasite was achieved in all cases when the inoculum contained 100 cysts or more. However, with 10 to 25 cysts, only a third of the patients became infected, and one or two weeks later cysts began to appear in the feces.[16] *In vitro,* shedding is achieved as long as there is humidity, prior exposure to pH 2 at a temperature of 37°C and sowing in a culture medium with a pH maintained at around 6.8 at the same temperature. These conditions mimic the passage of the cysts through the stomach and their hatching in the small intestine. In mice, disenrollment occurs in around 30 minutes.[16] In 1990, the first axenic cultivation of the parasite was recorded from isolates extracted from small mammals and, later, *in vitro* growth of the parasite from human material was reported. The modified TYIS-33 medium is the medium of choice for isolating, cultivating and axenizing it.[8]

1.3 Transmission

The normal route of infection for humans is ingestion of mature cysts, which can be transmitted by drinking untreated or poorly treated water (chlorine only); contaminated food (poorly washed fruit and vegetables), food can also be contaminated by cysts carried by insects; from person to person, through contaminated hands, in places where people gather, between family members when there is an infected individual and through contact with infected domestic animals. This last mechanism is still debatable, although the WHO (World Health Organization) considers giardiasis to be a zoonosis.[3,13,24] With regard to waterborne transmission, it is known that the majority of giardiasis epidemics in humans have occurred through groundwater contamination.[25] Contamination of the environment, including water, can be the result of human and animal action.[26]

1.4 Symptomatology, Pathogenesis and Parasite-Host Relationship

The first evidence of its pathogenicity was suggested by Fanthan and Porter (1916) when they correlated the parasite with the symptoms and diarrheal stools of soldiers returning from war.[27] Clinical manifestations of giardiasis range from asymptomatic infections to acute self-limiting diarrhea or chronic diarrhea with malabsorption, even in immunocompetent individuals.[28] The way in which this intestinal parasite causes damage to the host has yet to be fully elucidated.[29] The fact that most infections are asymptomatic has meant that the idea that this parasite is not pathogenic, being a commensal organism, has prevailed for many years.[30]

The mechanisms that lead to this variability are multifactorial, involving phenomena related to the strain of the parasite and the number of cysts ingested. The mechanisms related to the host are associated with its age, nutritional and immunological status and intestinal motility.[31] Only a minority of infections result in disease,[(32)) and] the majority of infections are therefore asymptomatic. In symptomatic individuals, there is diarrhea, steatorrhea, weakness, weight loss, nausea, abdominal pain, low-grade fever, and rarely the presence of mucus and blood in the stool. This acute form can last only a few days and its initial symptoms can be confused with those of diarrhea caused by medication, or of viral or bacterial etiology.[33,34]

The clinical manifestations of giardiasis have been associated with a wide range of clinical conditions, from mild, self-limiting enteritis to chronic, debilitating diarrhea with steatorrhea and weight loss. Although the infection is self-limiting in most healthy individuals, 30 to 50% can develop chronic diarrhea. The main complications in relation to chronic giardiasis are associated with poor absorption of nutrients such as vitamins, iron and lactose. These deficiencies rarely produce serious damage in adults, but in children, they can have serious effects and can compromise their physical and cognitive development.[5] The interval between infection and the appearance of symptoms is usually 2 weeks, but can last several months. The average duration of the disease in epidemic outbreaks is 66 weeks. In children, the symptoms most indicative of giardiasis are diarrhea with steatorrhea, irritability, insomnia, nausea and vomiting, loss of appetite (accompanied or not by weight loss) and abdominal pain.[35,36]

The mechanisms by which giardiasis causes diarrhea and intestinal malabsorption are not well known. When intestinal biopsies from infected individuals are examined using light microscopy, changes in the architecture of the mucosa can be observed. It can be completely normal or with partial or total atrophy of the intestinal villi. Even when the mucosa is morphologically normal, damage to the microvillus cells has been detected by electron microscopy.[3]

It is suggested that the morphological and functional alteration of the intestinal epithelium is a process that can trigger an inflammatory reaction by the parasite, due to the host's immune reaction, thus provoking a local anaphylactic reaction. This process can increase the renewal of enterocytes, leading to problems of malabsorption and diarrhea.[3] However, another factor suggested is the saturation of the mucosa by a large number of trophozoites, preventing the absorption of food. However, as the trophozoites are located more at the base of the villi, next to the crypts of Lieberkühn (Figures 6 and 7), there would still be a large free intestinal area through which absorption would take place.[37] In addition, symptomatic giardiasis may be associated with the growth of bacteria in the proximal portion of the small intestine.[36]

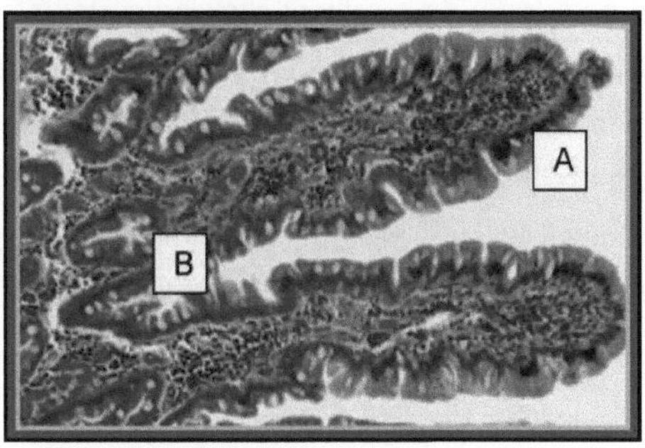

Figura 6. Small intestine (HE): (A) Most intestinal absorption occurs in the large portion of the intestinal villi. (B) In intestinal colonization, trophozoites are located more at the base of the villi, the so-called Lieberkühn crypts. Source: http://acd.ufrj.br/labhac/figura12.htm [38]

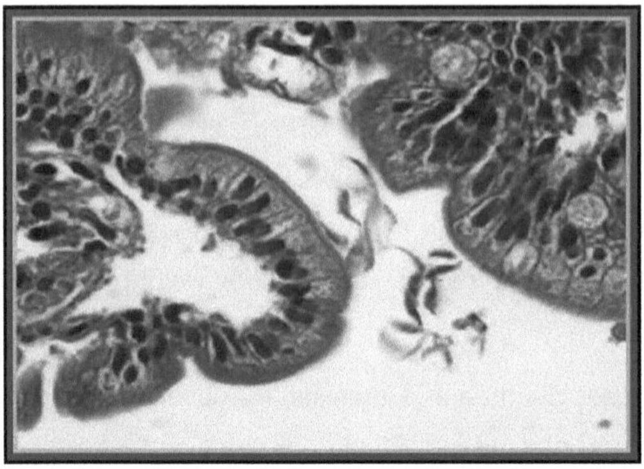

Figura 7. Small intestine (HE). *G. duodenalis* trophozoites colonizing the villi of the small intestine. Source: http://www6.ufrgs.br[(39)]

1.5 Laboratory Diagnosis

If giardiasis is suspected, direct examination of the stool can and should be carried out routinely. In diarrheal and recently eliminated stools, the search for trophozoites is indicated and in formed stools, cysts predominate, in cases of large excretion of cysts there is a time saving without prejudice to the reliability of the result.[(40)] When the result of the direct examination is negative, the presence of the parasite in the stool should continue to be investigated using concentration methods, such as the method of Faust *et al.* (1939) or Coprotest.[(41)]

A number of factors are responsible for the quality of the parasitological examination. In addition to the methodology used, it is important to consider the pattern of cyst excretion. Considering intermittent elimination, various approaches have been proposed to repeat the tests. As an individual can remain negative for up to 20 days without showing cysts, it is suggested that two to three fecal samples be examined every week for a period of 4 to 5 weeks.[(42)] In some patients with chronic diarrhea, the examination of several stool samples remains negative, despite the presence of trophozoites in the duodenum. In these cases, it is necessary to diagnose by examining the duodenal fluid and jejunal biopsy, which are very invasive methods. The most commonly used method for obtaining duodenal fluid is the "Entero-test."[(43)]

The immunological diagnosis of giardiasis by testing for *anti-Giardia* antibodies is an alternative methodology, but the cost still makes it difficult to use in routine laboratories.[(42)] However, serology is widely used in epidemiological surveys in various parts of the world. On the other hand, the detection of these antibodies has presented problems with the occurrence of false positives. In these reactions, IgG antibodies remain elevated for a long period of time, which makes it impossible to distinguish between past and recent infections. In this way, serological diagnosis can help in epidemiological surveys, but it has not shown adequate sensitivity and specificity for individual diagnosis [(44)].

Testing for coproantigens in feces has been widely used in both clinical and epidemiological studies, especially in cases where the clinical picture is suggestive and coproscopic methods are negative. These tests have satisfactory sensitivity and specificity[45] and the possibility of samples being preserved for a long time in formalin. The most commonly used antigen for detecting *Giardia* is GSA 65, a glycoprotein found in cysts or trophozoites.[46]

In recent years, molecular diagnostic techniques such as the polymerase chain reaction (PCR) or genetic probes that recognize the DNA or RNA of parasites have improved their detection in faeces. These methodologies are capable of identifying nucleic acids from both dead and live parasites, as well as being more sensitive and specific than microscopic and immunological techniques. However, the cost still limits these methodologies in the routine diagnosis of this parasitosis[(47,48,49)].

1.6 Aspects of the Epidemiology of Human Giardiasis

Giardiasis has a global distribution. In developing countries, rates of infection with this protozoan vary from 20 to 30%.[5] In some regions, rates can reach 68%[(51)] generating 50,000 new cases every year.[52] Around 20 million people in the world are infected with this protozoan and each individual can eliminate up to 900 million cysts a year.[53] This parasitosis is believed to be responsible for 2.5 million cases of diarrhea and nutritional deficiency in children in developing countries. [54] It is also estimated that 50% of infectious diarrheal diseases occur in children around three years of age who attend daycare centers. *G. duodenalis* is the most common recognized parasite causing intestinal infection.[2,52,55,56,57]

The prevalence of this parasite is higher in young humans, young animals and animals living in public shelters. Prevalence values also vary according to location and type of host.[58] The high prevalence observed in children may be due to a lack of hygiene habits, while in adults it seems that infection with this protozoan can confer a certain degree of resistance to subsequent infections. High prevalence rates are found in tropical and subtropical regions and among individuals with low economic status/[59]) and it is estimated that 50% of infectious diarrheal diseases occur in children attending day care centers.[60]

Studies have reported that 325 outbreaks associated with waterborne transmission of pathogenic protozoa have been documented, mainly in North America and Europe. Of these, 104 (32%) were associated with contamination of drinking water systems by the parasite *G. duodenalis*. The incidence of giardiasis in the United States is estimated at 2.5 million cases per year[(6)]1) Of the giardiasis outbreaks documented in the United States, 90% occur through water contaminated with infective cysts, and only 10% through food contamination. Some epidemics have been linked to contamination of water with cysts excreted by animals such as beavers and muskrats.[(62)]

In Brazil, a wide variation in the frequency of this protozoan has been observed.[(45,63,64,65)] An evaluation of the prevalence of enteroparasitosis in schoolchildren in Paranà showed that 50.73% of infections were due to *G. duodenalis* .[(69)] On the other hand, in the municipality of Criciùma, State of Santa Catarina, there was a low incidence of 4.3%.[(64)] In children hospitalized with diarrhea in Goiânia, State of Goiàs, 9.9% of the stools analyzed were positive for this protozoan/[70]) Carvalho-Costa *et al.* (2007) estimated the frequency of intestinal parasites in 218 children with acute diarrhea and severe dehydration, admitted to a pediatric hospital located in Rio de Janeiro, *G. duodenalis* was the second most frequent parasite, found in 4.7% of the samples.[(71)] Volotao *et al.* (2007), in their study involving 310 children from a low-income area also located in Rio de Janeiro, observed the prevalence of intestinal parasites in 60% of the samples analyzed, of which 27.7% were positive for *G. duodenalis* cysts. The highest level of infection with this protozoan was observed in children aged between two and four years.[72] Researchers assessed the prevalence of this parasite in a study carried out in the state of Minas Gerais and found 26.3% of children infected with this parasite, whose infection was associated with socioeconomic and environmental factors.[73] In addition, a study investigated the presence of *G. duodenalis*[(45)] in children from a nursery school on the outskirts of Belém and detected its presence in 26.9% of the samples.

In the state of Sao Paulo, this panorama has not changed. In the municipality of Presidente Bernardes, when analyzing stool samples from children attending nursery schools, a prevalence of 16% was detected. Analysis of the feces of these children's relatives confirmed that some mothers and siblings were parasitized and had this protozoan in their feces.[(74)] In the 1990s, an epidemiological survey of children in the municipality of Mirassol, in the northwest of São Paulo, showed a prevalence of 61.1% of this parasite.[(75)] In Botucatu, a study carried out in five municipal nurseries showed that 23.7% (2002) and 21.4% (2003) of individuals had *G. duodenalis* in their feces.[(76)] In the same municipality, in another study carried out with the same objective, a survey of enteroparasitosis, three daycare centers of the municipal network were studied, in the rural area, in the center of the city and the other on the outskirts, the fecal analyses showed results of 69.6%, 52.7% and 69.6% for the parasite in question, respectively. Once again, its parasitic prevalence is shown, with the highest percentage due to the analysis of three fecal samples from each child over a seven-day interval, followed by *Ascaris lumbricoides* (20.4%) and *Trichuris trichiura* (19.0%). In Campinas, the most prevalent parasite in children was *Ascaris lumbricoides* (12.3%), followed by *G. duodenalis* (8.2%). In this study, all the parasites found showed a percentage of 30.6%, with only the two presented showing a frequency of more than 20%.[(78)] In another study, in the municipality of Assis, in a total of 18,366 parasitological stool tests, 32.3% of these were diagnosed with parasites, with *G. duodenalis* being the most prevalent, with a frequency of 8.7%.[(79)]

The importance of enteroparasites as colonizers of the gastrointestinal mucosa of HIV-1 carriers is unquestionable.[(33)] *G. duodenalis* is among the most frequently cited protozoa.(80,81,[82]) An investigation of HIV-1 patients from the capital of the state of São Paulo showed a significant association between the manifestation of diarrhea and the presence of giardiasis.([83]) In the municipality of Sao José do Rio Preto, in the north-west of São Paulo, three studies have been carried out in this regard.[(33,34,84)] The frequency of *G. duodenalis* was 4.04% and 9% and an association between its presence and diarrhea was only found in the adult population.[(34)]

1.7 Genotypes of the Agents Causing Giardiasis

In 1979, the World Health Organization (WHO) considered giardiasis to be a zoonosis. Molecular studies have shown that *G duodenalis* is a complex species, made up of at least seven distinct genotypes. Genotypes A and B have been linked to infections in humans and have also been detected in some mammals. In contrast, the other genotypes have shown preferences among different animal species. Genotypes C and D include isolates from dogs, which seem to be specific to them, genotype E is related to production animals, such as pigs, sheep and cattle, and genotypes F and G seem to be exclusive to domestic cats and rats, respectively.[(85,86,87,88)] However, the zoonotic potential of the disease is still debated, particularly among domestic animals.[(89,90)] In addition, new genotypes are still being described.[(91,92)]

Based on knowledge of the frequency of the parasite in different species of animals and humans and its genotypes, four transmission cycles are responsible for maintaining the parasite in the hosts (Figure 9). One of the cycles has dogs and cats as its main hosts, the other humans, the third cattle and the fourth wild animals. There is the possibility of transmission between hosts of the same cycle and from one cycle to another directly or via water. The frequency of transmission is unknown for each cycle.[86,93]

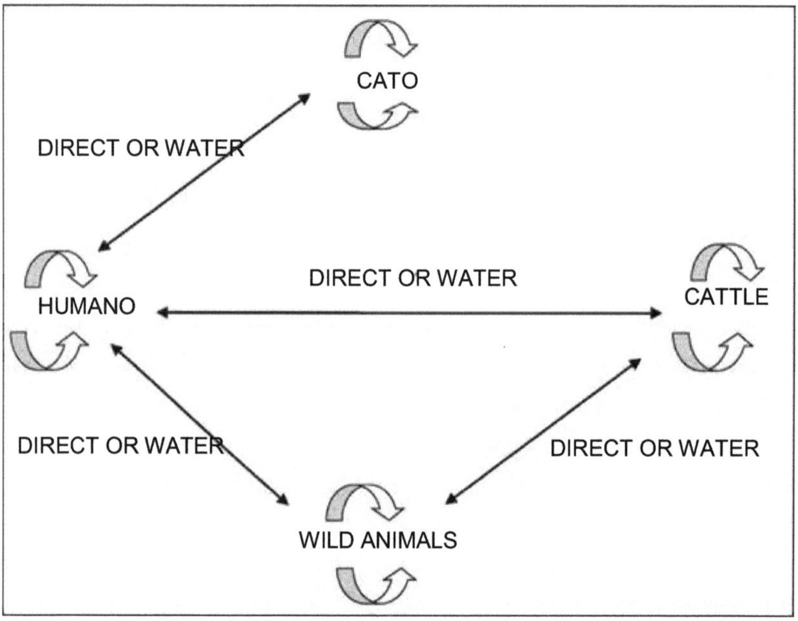

Figura 8. The four main transmission cycles of *G. duodenalis.* Adapted(93)

Most studies have carried out analyses that allow the identification and comparison of *G. duodenalis* isolates by different gene targets.[87,94,95,96] One of these is the *β-giardin* gene, which has been shown to be a good parameter for identifying subgenotypes, which are extremely important for identifying the source of infection in outbreak situations.[89,97,98]

Group or genotype A consists of two subgenotypes, A1 and A2. The A1 subgenotype consists of parasites isolated from humans and animals, which has been the subject of study in analyzing the potential for zoonotic transmission of giardiasis. A2 comprises parasites isolated from humans,(⁹9) however, it has also been detected in animals. Genotype B was thought to be exclusive to humans, however it has been described in dogs, beavers, cattle, horses, monkeys, rats and sheep, and associated with severe clinical symptoms such as malabsorption, decreased weight gain, foul-smelling and malformed stools.[87] Genotype B has been detected in human feces in Argentina,[95] the United Kingdom,(¹⁰0) Italy(⁹⁷) and France.[101] On the other hand, studies carried out in Portugal-¹⁰²) and Mexico[89] show only the detection of genotype A in human fecal samples. In addition, multiple infections have been detected in some studies around the world.[89,100,103,104,105]

To date, three surveys carried out in Brazil, two in the Southeast and one in the South, have shown discordant results regarding the distribution of *G. duodenalis* genotypes in human and non-human fecal samples. Initially, in the municipality of Rio de Janeiro, southeastern Brazil, the A1 and A2 subgenotypes were detected in human fecal samples. In addition, A1 was the only one found in the domestic animals investigated.[72] Furthermore, a case of *G. duodenalis* infection associated with a child and his dog and both isolates were characterized as A1, suggesting that despite the low incidence, this data indicates a possible existence of the zoonotic cycle carried out by this parasite. Subsequently, in the south-east of the state of São Paulo, genotypes A2 and B were detected in human feces, genotypes A1 and F in cat feces and genotypes C and D were detected in canine fecal samples.[94] Recently, Volotâo and colleagues (2008) detected only genotype A1 in monkeys in the state of Santa Catarina, in southern Brazil.[106]

Molecular analyses show that the genetic distance shared by genotypes A and B may exceed that used to describe other protozoan species.[2]

In studies carried out in Canada and Australia, it was shown that a small proportion of a herd (less than 20%) harbored *G. duodenalis* genotype A.[13] In France, Bertrand and Schwartzbrod (2007), when comparing the presence of genotypes A, B and E in wastewater samples from two municipal treatment areas and a slaughterhouse, found genotypes A and B in the first type of sample and A and E in the second. They concluded that the combination of genotypes A and B seemed to indicate contamination of human origin while that of A and E of bovine origin.[107] The determination of aquatic molluscs parasitized by *Giardia duodenalis* of genotype A was suggestive of contamination of the water with human faeces.[108] Data such as this can serve as a biological indication of water contamination through parasite cysts, and can thus determine water quality. In an outbreak of *giardiasis* in Norway in 2004, it was suspected that the source of infection was waterborne, where molecular characterization of this source identified the A genotype of *G. duodenalis*. Curiously, this suspicion was discarded after the molecular characterization of the individuals with the disease showed parasites corresponding to genotype B.[109] In Hungary, analysis of the genotypes in surface water and sewage samples revealed genotype A in seven samples, B in one sample and mixed infection in four samples.[110]

Another aspect to be considered is the association between these genotypes and the clinical picture. Evidence from isolates of this parasite in Australia shows that genotype A differs from genotype B in terms of virulence. It was observed that children parasitized by the first genotype had a higher frequency of diarrhea.[111,112] In agreement, Sousa *et al.* (2006) and Sahagùn *et al.* (2008), reported that genotype A seems to be associated with gastrointestinal complications, thus presenting symptomatic infections in isolates from Portugal and Spain, respectively.[102,113] In contrast, a study carried out in London also found that genotype B is associated with persistent diarrhea in humans.[100]

1.8. Justification

As we have seen, the distribution of *G. duodenalis* genotypes shows considerable geographical variation. Genotype A is predominantly represented in some endemic areas, while genotype B is more frequent in other regions of the world.[94,112] In addition, multiple infections have been detected in some studies,[97,100,103] a fact that may have occurred due to waterborne contamination,[104] but also due to a specific epidemiological situation in each area.[105] Another hypothesis to be considered is that this variability in the distribution of genotypes may be influenced by the parasite's interactions with its hosts in different areas around the world.

In Brazil, there is a wide variation in the frequency of this protozoan parasite and in some regions of the state of Sao Paulo, up to 40% of individuals can harbor this parasite.[33] Only two genotyping studies of *G. duodenalis* in humans have been carried out in Brazil and discordant results were observed when comparing the frequency and genotypes of isolates detected in the municipality of Rio de Janeiro and between localities in the southeast of the state of Sao Paulo. Therefore, the imminent risk of zoonotic transmission among the inhabitants of northwestern São Paulo prompts us to carry out a broader investigation, including its association with clinical manifestations and the participation of the human host as a source of infection for the community. In addition, the wide diversity of geographic, social, economic and climatic characteristics in Brazil has been reported as critical factors in modulating the frequency of different enteroparasites in various studies relating etiological agents of diarrhea.[33,64,83]

2. MATERIAL AND METHOD

2.1. . Study Area

Fecal samples from individuals living in two municipalities in the interior of the state of São Paulo were analyzed. The first, the city of Araçatuba (latitude 21°12'32" and longitude 50°25'58", with an altitude of 390 m from sea level) is located 524 km from the state capital. It has a population of 182,204 inhabitants living in an area of 1,167 km^2. The second municipality where the research was carried out was Sao José do Rio Preto (latitude -20°89'72" and longitude -49°37'44", with an altitude of 489 m from sea level), also located in the northwest of São Paulo, 450 km from the city of São Paulo. Its population of 419,632 lives in an area of 431 Km^2. (115)

2.2. . Fecal Samples

From July 2007 to March 2008, fecal samples from individuals who tested positive for this protozoan by parasitological examination were screened. A total of 4,300 samples were analyzed at the Central Laboratory of the Hospital de Base (HB) in Sao José do Rio Preto. HB is a tertiary hospital that serves a large population in the northwest of the state of São Paulo. 150 individuals were selected who had *G. duodenalis* in their feces, and these results were obtained by means of a parasitological examination of feces. A fecal sample was collected from each of the 145 individuals who agreed to take part in the study. In addition, five stool samples from HIV-1 carriers positive for this enteropathogen, already in the Microorganism Research Center (CIM) of the Sao José do Rio Preto Medical School (FAMERP), were included.

During the months of July to December 2009, fecal samples were collected from two daycare centers located in the Alvorada neighborhood, from individuals living in the Sao José neighborhood and from employees of the Zoonosis Center in the municipality of Araçatuba. Fecal samples, with or without diarrhea, from 134 children aged 4 months to 6 years and 20 adults were included, totaling 154 samples. After a detailed explanation of the objectives of the study and the participants signing an informed consent form, a stool sample was collected from each individual. This sample was placed in two containers; one containing 10% formaldehyde for parasitological analysis, which was stored at room temperature, and the other without the addition of a preservative. This sample was stored at -20°C and used for molecular analysis.

All the samples collected in Araçatuba were sent directly to the CIM at FAMERP for parasitological and molecular analysis. The samples collected at HB were sent to the CIM for confirmation of the coproscopic diagnosis and the samples that were positive by microscopic examination were sent to the Interdisciplinary Medical Research Laboratory of the Oswaldo Cruz Institute - Oswaldo Cruz Foundation, Rio de Janeiro, for genotyping of the *G. duodenalis* isolates. The work was approved by the Human Research Ethics Committee (process 3456/2007) and the Animal Experimentation Ethics Committee (process 109508/2007) of FAMERP (Appendices 1 and 2).

2.3. Parasitological Diagnosis

Human fecal samples were examined in order to detect the presence of the parasite using optical microscopy. The techniques used were centrifugation with 33% zinc sulphate [116] and spontaneous sedimentation using the Hoffmann, Pons & Janer method [45] and identification of the parasite using 10x and 40x objectives.

2.4. Molecular analysis

2.4.1 DNA extraction

The parasite's genetic material was extracted and purified from two hundred microliters of *Giardia duodenalis-positive* fecal suspension using the Hoffman, Pons & Janer method.[45] The process was carried out using the Qlamp DNA Stool Mini *Kit* (Qiagen, Germany), following the manufacturer's instructions, with the exception of the volume of AE buffer used to elute the DNA, which was reduced to 80 µL or 100 µL and the lysis temperature which was increased to 95°C, the latter procedure also suggested by the manufacturer.

After the DNA extraction process, the samples were quantified using spectrophotometry (NanoDrop 1000 Spectrophotometer, USA) with an absorbance of 260nm. Those with a DNA concentration below 50 ng/µL were subjected to vacuum centrifugation (VacufugeTM, USA) in an estimated time of around 30 minutes. In cases where the DNA concentration did not reach the desired values (above 50 ng/µL), it was necessary to repeat the process for a given sample at a local temperature of 30°C for 15 minutes.

2.4.2 Genotyping of *Giardia duodenalis* isolates

Samples from Sao José do Rio Preto

PCR amplification was carried out as previously described.[117] The 753 bp fragment amplified is part of the gene coding for the *β-giardin* protein. *The* amplification was carried out using the G7 primer oligonucleotide as a primer and the G759 reverse primer oligonucleotide. A second amplification was carried out to obtain a 511 bp fragment of the same gene amplified using the BGIA1F and BGIA1R primer oligonucleotides.[(89)]

The PCR was carried out in a final volume of 50 μL, containing 1X buffer, 1.5 mM $MgCl_2$, 200 μM of each dNTP, 10 μM of each primer oligonucleotide, 2.5 U of AmpliTaqGold polymerase enzyme (Applied Biosystems, USA) and 1-5 μL of purified DNA. The reaction was carried out in a thermal cycler (Applied Biosystems, USA) under the following conditions: 94°C for 5 min; 35 cycles of 30 s at 94°C, 30 s at 65°C, 1 min at 72°C; followed by a final extension of 7 min at 72°C.

Aliquots of 10μL of the amplified products were applied to a 1.2% agarose gel (Life Technologies) in 0.5x TBE buffer (89 mM Tris base, 89 mM boric acid and 2mM EDTA) and subjected to electrophoresis in the same buffer at 100V for approximately 1 hour. A 100pb molecular size marker (Invitrogen, USA) was used. The gel was stained with 0.5μg/mL ethidium bromide (Invitrogen, USA) for 30 min and destained with distilled water for 2 h. After electrophoresis, the gel image was captured using a photodocumentation system (GelDoc, BioRad, Hercules, CA, USA).

As standards, we used trophozoites *of G. duodenalis*, strain WB (ATCC 50803), previously characterized as genotype A1, grown axenically in TYI-S-33 medium enriched with 10% fetal bovine serum and 0.1% bovine bile for 36 h at 37°C in a 5% CO_2 atmosphere,[118] kindly provided by the Herta Meyer Laboratory, Carlos Chagas Filho Biophysics Institute - UFRJ. As a control for genotype B, we used DNA extracted from isolate GM-1 (NCCID-CDC, Atlanta-GA, USA). As negative controls, we used DNA extracted from axenic cultures of *Entamoeba histolytica*, *Trichomonas vaginalis* and clinical samples positive for *Cryptosporidium parvum.*

In order to detect single nucleotide polymorphisms (SNPs), all the amplification products obtained from the second amplification were sequenced in triplicate and in both directions. The amplified products were purified using the Wizard® SV Gel and Pcr Clean-Up System kit (Promega, Madison, WI, USA) and sequencing was carried out using the ABI PrismTM v 3.1 *Dye Terminator Cycle Sequencing Kit* (Applied Biosystems, Foster City, CA). Where 1 μL of the *dye Terminator mix*, 3.2 μM of the primer oligonucleotide (BGIA1F or BGIA1R) and 2 μL of the purified PCR product were used in each reaction with a final volume of 10 μL. The reaction conditions were: 40 cycles of 94°C/10 s, 50°C/5 s and 60°C/4 min.

The sequences obtained were analyzed using the Chromas program (http://www.technelysium.com.au/chromas.html) (25/09/2005) and compared with the β-giardin sequences of *G. duodenalis* already described in *GenBank* using the Clustal W[(55)] algorithm in MEGA version 3.1 ()http://www.megasoftware.net) (22/05/2007. The phylogenetic analysis of the β-giardin *locus* was based on the consensus alignment of approximately 350 bp obtained from Clustal W. The phylogenetic analysis was developed in MEGA and the distance estimate used was the JIN & NEI[119] equation (model

Kimura 2-parameter). The phylogenetic trees were constructed using the neighbor-joining algorithm.[120] For each construction, the veracity of each branch was checked by *bootstrap* analysis (1000 repetitions). The sequences used as references were those belonging to the main genotypes of *G. duodenalis* and were aligned with the sequences obtained from the new isolates.

Araçatuba samples

The PCR procedure for amplifying a fragment of the *β-giardin* gene was carried out using a semi-nested PCR protocol as described by Caccio *et al.* (2002).[117] The first PCR reaction yielded a 753pb fragment. The oligonucleotide primers used for this amplification were G7 (AAGCCCGACGACCTCACCCGCAGTGC) and G759 (GAGGCCGCTGGATCTTCGAGACGAC). The second reaction used the G759 oligonucleotide from the first reaction and the G376 oligonucleotide (cATAAcGAcGccATcGcGGcTcAGGAA). The fragment obtained was 384 bp.

A final volume of 50 µL was made for PCR, containing 1X buffer, 1.5 mM $MgCl_2$, 200 µM of each dNTP, 10 pmol of each oligonucleotide primer, 2.5 U of Taq polymerase enzyme (Fermentas Life Sciences - Sinapse Biotecnologia, Brazil) and 5 µL of purified DNA. The reaction was carried out in a thermal cycler (Techno) under the following conditions: 94°C for 5 min; 35 cycles of 30 s at 94°C, 30s at 63°C, 1 min at 72°C; followed by a final extension of 7 min at 72°C. For the second reaction, 1 µL of the product from the first reaction was used for the same amplification conditions.

Products from both PCRs were subjected to electrophoresis in a 1.5% agarose gel at 110V for 50 minutes. The molecular size standard was 100pb. The gel was stained with ethidium bromide, visualized by transillumination with UV light and its image digitized (Transilluminator FBDLT-88). *G. duodenalis* trophozoites were used as standards, strain WB (ATCC 50803), kindly provided by the Herta Meyer Laboratory, Carlos Chagas Filho Biophysics Institute - UFRJ. As negative controls, we used DNA extracted from axenic cultures of *Entamoeba histolytica*, *Trichomonas vaginalis* and a clinical sample positive for *Cryptosporidium parvum*.

An analysis of restriction fragment length polymorphism (RFLP) using products amplified by PCR and semi-nested PCR was carried out as described by Caccio *et al.* (2002).[117] Aliquots of 10-12 µL of the products from the first PCR reaction were digested with the addition of 10

U of *Hae III*. For the products obtained from the second reaction, the *Hha I* enzyme was used. Both procedures were carried out using a final volume of 20 µL for 4 hours at 37°C.

The first digestion produces fragments of 202, 201, 150, 126 and 74 bp, characteristic of genotype A, and fragments of 202, 176, 150, 117, 84 and 24 bp for genotype B. The second digestion used to distinguish the A1 subgenotype from the A2/A3 subgenotypes The A1 subgenotype of *G. duodenalis* was cleaved into fragments of 193, 104, 70 and 17 bp, while those belonging to the A2 / A3 genotypes gave rise to fragments of 210, 70 and 34 bp[72]

The products obtained after the enzymatic digestion process were analyzed using 2% agarose gel electrophoresis, 110V for 1 hour. The molecular size standard was 50pb. The gel was stained with ethidium bromide, visualized by transillumination with UV light and its image digitized (Transilluminator FBDLT-88).

2.5. Statistical Analysis

Analyses were carried out using the statistical software R version 2.4 (Copyright 2006 The R Foundation For Statistical Computing ISBN 3900051-07-0). Fisher's exact test and the chi-squared test were applied to obtain independence between the proportions, with a significance level of $p<0.005$.

3. RESULTS

3.1. Detection of *Giardia duodenalis* using optical microscopy

In the municipality of Sao José do Rio preto, the samples showed a frequency of 3.5% (150/4,300), all of which were diarrheal. Of the samples analyzed, only 83 confirmed parasitological diagnosis for the parasite studied.

The samples from the municipality of Araçatuba showed a frequency of 25.3% (39/154) for this protozoan,26.8% (36/134) in children and 15% (3/20) in adults. Only two fecal samples with the parasite were diarrheal (5%).

3.2. Genotyping of *Giardia duodenalis* isolates

Samples from Sao José do Rio Preto

Of the samples analyzed, only 83 confirmed the parasitological diagnosis and only 55 (66.2%) of these were amplified by PCR (one sample was not diagnosed because it did not show good results in its sequencing). Analysis of the sequences of the 511 bp gene fragment (Figures 9 and 10) obtained from these amplified samples showed that all the *G. duodenalis* isolates were genotype A, with 38 being A1 subgenotypes and 16 A2 subgenotypes. Genotype B was not found in the samples analyzed in this study.

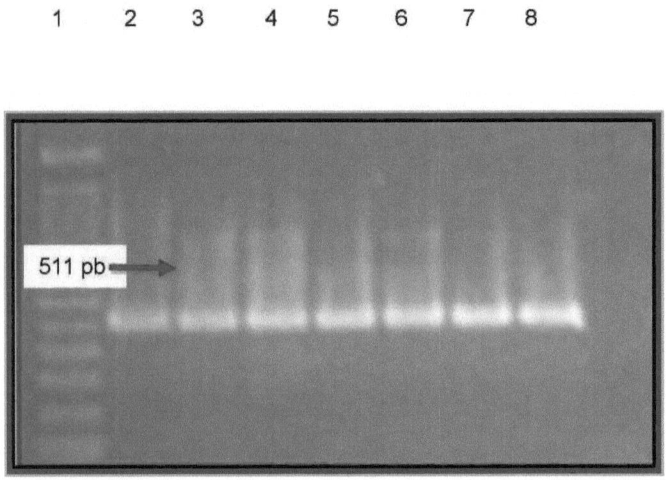

Figura 9. Agarose gel at 1% concentration with PCR products from the 2nd reaction. (1) 100 bp molecular weight marker. (2, 3, 4, 5, 6, 7) 511 bp fragment of the *β- giardina* gene. (8) Negative control.

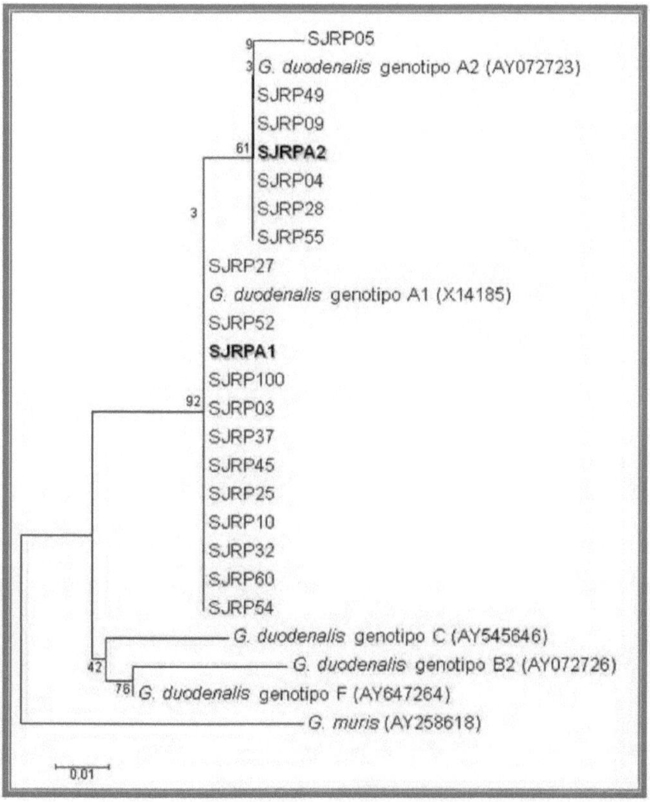

Figura 10. Phylogenetic tree of G. duodenalis isolates based on the β-giardin gene sequences of isolates deduced by the Neighbor-Joining algorithm using Kimura 2-parameters. *GenBank* sequences are indicated by the accession numbers in brackets. SJRPA1 and SJRPA2 correspond to the samples analyzed that had sequences identical to the A1 and A2 genotypes, respectively.

Araçatuba samples

Of the 39 parasitologically positive samples for *G. duodenalis* analyzed from Araçatuba, 32 (82%) showed conclusive molecular results. The first semi-nested PCR reaction amplified a 753 bp fragment (Figure 10). The second reaction of this protocol amplified a 384 bp fragment (Figure 11).

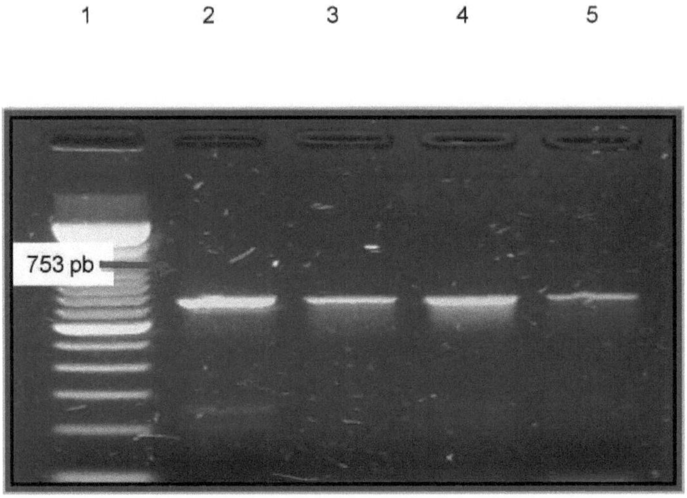

Figura 11. Agarose gel at a concentration of 1.5% with PCR products from the 1st reaction. (1) 100 bp molecular weight marker. (2, 3, 4, 5) 753 bp fragment of the *β-giardin* gene.

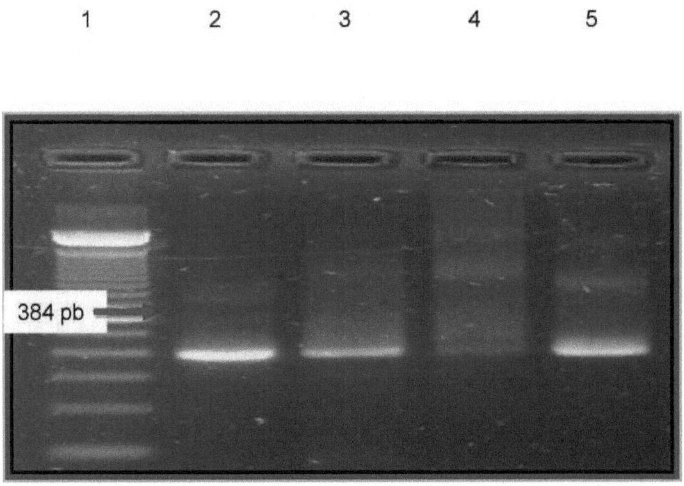

Figura 12. Agarose gel at 1.5% concentration with semi-nested PCR products, 2nd reaction. (1) 100 bp molecular weight marker. (2, 3, 4, 5) 384 bp fragment of the *β-giardin* gene.

The PCR product from the first reaction was digested with the *Hae* III restriction enzyme and showed fragments of 202, 201, 150, 126 and 74 bp. The results obtained at this location identified 32 samples belonging to genotype A and seven isolates that did not show amplification. No samples with genotype B were detected. The 384 bp fragment of the *β-giardin* gene obtained in the second PCR reaction was digested with *Hha* I. The presence of fragments of 193, 104, 70 and 17 bp were detected. The A1 genotype was identified in 10 isolates, while the A2 genotype was found in 22 samples with fragments of 210, 70 (two bands found) and 34 bp (Figure 12).

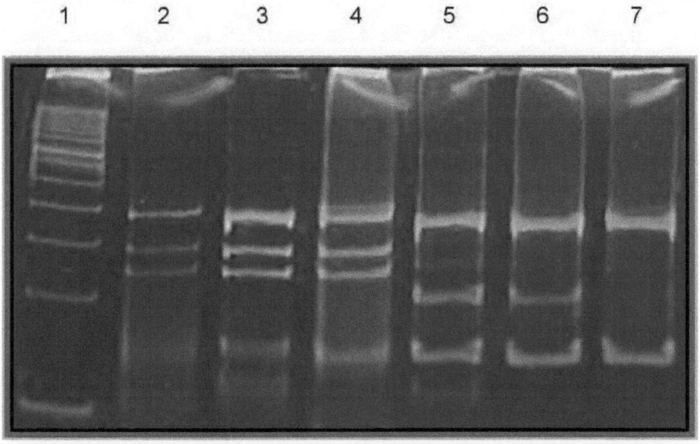

Figura 13. Polyacrylamide gel.(1) 50 bp molecular pattern. (2, 3, 4) Enzymatic digest with *Hae* III restriction endonuclease showing genotype A. (5, 6) Enzymatic digest with *Hha* I restriction endonuclease showing subgenotype A1. (7) Enzymatic digest with *Hha* I restriction endonuclease showing subgenotype A2.

3.2 Correlation between *G. duodenalis* Genotypes *and* Fecal Aspect in the Municipalities of Sao José do Rio Preto and Araçatuba

The samples from Sao José de Rio Preto showed no correlation between the presence of the genotype and fecal appearance. 70.3% had the A1 subgenotype and 29.6% had the A2 genotype.

With regard to the samples from Araçatuba, we found that 93.7% were non-diarrheal. Of these, nine were identified as subgenotype A1 (28.1%) and 21 as subgenotype A2 (65.6%). Only two samples had a diarrheal appearance, one with the A1 subgenotype and the other with the A2 subgenotype. No significant results were observed between stool consistency and the subgenotype detected (Table 1).

Table 1. Association between the presence of *Giardia duodenalis* genotypes *and* fecal appearance in samples from a human population in the municipalities of Sao José do Rio Preto and Araçatuba, northwestern São Paulo State, Brazil.

Sample		Fecal appearance	
Area	Genotypes detected	Not diarrheic	Diarrheal
Araçatuba (n = 32)	A1 (n= 10)	09 (28,1%)	01 (3,1%)
	A2 (n= 22)	21 (65,6%)	01 (3,1%)
Sao José do Rio Black (n = 54)	A1 (n= 38)	0	38 (70,3%)
	A2 (n= 16)	0	16 (29,6%)

4. DISCUSSION AND CONCLUSION

The southeastern region of Brazil is home to 46.5% of the Brazilian population, and the state of Sao Paulo has the vast majority of this, with more than 40,000,000 inhabitants[114]. In this study, we will present, for the first time, information on the genotypic frequency of *G. duodenalis* in humans in the municipalities of Araçatuba and Sao José do Rio Preto, in the northwest of São Paulo. The data presented contributes to future regional and national comparisons, as well as to other developing and industrialized countries, offering subsidies for understanding its epidemiology and the development of measures for the prevention and control of giardiasis.

In Brazil, there is great variation in the frequency of *G. duodenalis*. An evaluation of enteroparasitosis in schoolchildren in Paranà showed that 50.73% of infections were due to this parasite,[69] but in Santa Catarina, in the child population, there was a low incidence of 4.3%.[64] This variation is observed in various regions of the country[75,78]. In the state of São Paulo, it was described in Presidente Bernardes,[74] Mirassol[75] and Campinas,[78] where the frequency of this protozoan was 16%, 61.1% and 8.2%, respectively. Our results are not dissimilar to the national panorama, which showed a variation in frequency from 3.5% in São José do Rio Preto to 25.3% in Araçatuba. However, in this study, a fecal sample was collected from each individual, which may have influenced this sample, due to the intermittent release of the parasitic forms of this parasite.[72] On the other hand, it is known that giardiasis is more frequent in children than in adults, a fact observed in the samples collected in Araçatuba, where the majority of the population studied were children (87%), which is in line with the literature.[69,76]

Molecular studies have shown that *G duodenalis* is a complex species, made up of at least seven distinct genotypes. Genotypes A and B have been linked to infections in humans and have also been detected in some mammals. On the other hand, the other genotypes have shown preferences among different animal species. [85, 86, 87] Interestingly, in this study only genotype A and its respective subgenotypes A1 and A2 were detected and no samples showed genotype B. In the southeast of the state of Sao Paulo, 450 km from Sao José do Rio Preto, [94] genotype B was detected in human feces. In other locations, such as Argentina,[95] the United Kingdom,[100] Italy[97] and France,[101] this same genotype was found. However, studies carried out in Portugal,[102] Mexico[89] and Rio de Janeiro (850 km from the north-west of São Paulo), also in the south-east of Brazil, showed only the detection of genotype A in human fecal samples.[72]

In addition to genotyping *G. duodenalis* in humans, the importance of this process in animals, especially mammals, should also be emphasized, as these are also parasitized by the same genotypes found in humans (A and B). However, the zoonotic potential of giardiasis is still debated ((90,97)).

Studies carried out in India/[105]) Australia^[11]) and Brazil([72]) detected genotypes A and B in dogs, while another study revealed these genotypes in cats in Brazil.([94]) In addition to domestic animals, these genotypes have been found in other animals, such as beavers, cattle, horses, monkeys and sheep.[(87)] Also in Brazil,[(72)] a case was reported where genotype A1 was present in both the child and his dog.

In addition, it is well known that waterborne transmission is one of the main routes of transmission of parasitic diseases[(120)] and that this can include the participation of domestic animals, wild animals and humans, which has been demonstrated in Hungary' where the presence of genotypes A and B was found in contaminated water.([110])

In this study, the absence of genotype B in the two municipalities may have been due to the fact that not all the cysts detected by the coproscopic method were identified by molecular methodology; this has already been suggested in the literature.[(94)] Furthermore, as previously mentioned, it is known that the cysts of this parasite are eliminated intermittently in the feces and that fecal material can contain DNA polymerase inhibitors, which can compromise the final amplification result. [(72)] Reinforcing this idea, of the samples positive for the parasite in the municipality of São José do Rio Preto, 66.2% were amplified by PCR (55/83), and 28 samples were inconclusive by the molecular method. In Araçatuba we were more successful in terms of coproscopic and molecular positivity. In this city, 82% of the positive samples were amplified (32/39), so 7 of them did not show conclusive molecular results.

In recent years, a lot of information has emerged about the cellular and molecular biology of *G. duodenalis*. This is probably just the beginning of a period of intense research. At the moment, isolates of this protozoan presenting different genotypes are being sequenced.[(121)] However, it is necessary to establish what would be the appropriate criteria for describing new genotypes and a standardization related to the use of molecular markers. These definitions would prevent certain divergences in the results obtained through molecular research. Confirming this fact, Robertson *et al.* (2007),[(109)] after carrying out the molecular characterization of isolates of G. duodenalis isolates in human fecal samples by the *gdh, β giardin and tpi* genes, they obtained some sequences that showed differences of up to four single nucleotide polymorphisms (SNPs) in relation to the genotype sequences previously described, although they did not designate another subgenotype with their results, citing

Lalle *et al.* (2005),[97] who described a new genotype based on a single SNP. Another problem was raised by the same authors, where the same isolate could fall into different genotypes if classified by more than one gene. In relation to the findings, these authors attributed the possibility of the presence of more than one isolate in the same sample, but without confirmation of the fact, although this is the most accepted explanation.

In both municipalities, the A1 subgenotype (55.8%) was the most prevalent, compared to the A2 subgenotype (44.1%). These data are consistent with the literature on their prevalence in human isolates [72,89] However, when we compared their frequency by location, we found that in Araçatuba the A2 subgenotype was more prevalent (68.7%), unlike Sâo José do Rio Preto (29.6%). This fact was also found in Bangladesh,[32] Mexico,[122] the south-eastern region of Brazil[94] and European countries.[90] Therefore, based on these data and our results, we can infer that the geographical distribution of these subgenotypes has different patterns around the world.

Diarrhea can be infectious or non-infectious, the former being associated with bacterial, viral and parasitic agents.[123] Contradictory results of this association have been observed in different locations.[124] In Brazil, this profile is quite similar, where *G. duodenalis* has been shown to be an important pathogen for this intestinal alteration.[34] In this sense, we sought to verify some correlation between the genotypes detected of this protozoan and the fecal aspect of the samples. Studies in India[125] have shown that genotype B is more associated with diarrhea. In contrast, genotype A was shown to be more associated with diarrheal stools in Bangladesh/[32] In the municipality of Sao José do Rio Preto, no association between genotypes and diarrhea could be verified, since all the samples presented this fecal aspect. However, the fact that all the individuals had diarrhea confirms previous evidence that genotype A is more virulent than genotype B.[111] In the municipality of Araçatuba, no significant association was observed between diarrhea and the genotypes found. Although two diarrheal samples positive for this parasite had the A1 and A2 subgenotypes, respectively.

For the parasite to be successful in colonizing the host's intestinal tract, thus promoting infection, experimental studies[126] show a strong dependence on the parasite-host relationship. The factors that may be related to the susceptibility of individuals to being colonized by *G. duodenalis* relate to their nutritional status, age and immune response. The main parasite factors in this relationship are the role of the adhesive disk, flagella, and variant surface proteins (VSPs) that function as receptors.[127]

Several studies are being carried out in an attempt to elucidate the pathogenic mechanism of giardiasis. Based on these, it is understood that this disease occurs through a multifactorial process and, therefore, further conclusions are needed on the interaction between the parasite and the host.(121) Greater understanding of this interaction could help in understanding the genotypic distribution of the parasite. Evidence of host variability, behavior of the intestinal mucosa and immune response, in association with the genotypes of this parasite distributed in different areas, could offer more objective conclusions about the epidemiology of this protozoan, as well as the clinical characteristics of this parasitosis.

The results obtained in this study, together with the knowledge acquired from other studies, lead us to some considerations. The uniformity of genotypes in the northwest of São Paulo can probably be explained by their origin, since the entire population studied lives in close geographical locations. It can also be assumed that the great variation in social, climatic and geographical characteristics in Brazil, which has been reported in various studies on etiological agents as important factors that adjust the frequency of different pathogens,[(33)] which may explain the difference observed between the present results and those found by Souza et al. (2007). Another hypothesis would be that, since the Brazilian population has a very heterogeneous ethnic composition, the result of the miscegenation of numerous native indigenous populations and immigrants from Europe, Africa and Asia, and that the flow of immigration was not uniform in the different regions of the country, it can be concluded that the distribution of these genotypes may be related to the parasite-host relationship in various endemic areas of the country.

In line with this reasoning, studies show that the polymeric immunoglobulin receptor (pIgR), conserved in mammals, has an important evolutionary function in vertebrates. This receptor transports IgA and IgM and is highly expressed in the intestine. However, there is still little direct evidence of its importance in defense against enteric pathogens, but a recent study has shown that this receptor appears to play a critical role in the immune response against *G. duodenalis*.[((128))] Another study shows the ability of this protozoan to interfere in the host's immune response by inhibiting the production of interleukins by dendritic cells.(129) Furthermore, a combination of mechanical forces and lectins from surface of this pathogen may be involved in the specific recognition of human enterocytes(130) and the release of certain proteins by , deactivating host immune factors such as nitric oxide.

Recently, in order to evaluate the behavior of the intestinal epithelium, investigations of patients with chronic giardiasis have shown that there is an induction of apoptosis by enterocytes during the progression of the disease[(131)]. These results indicate several mechanisms involved in the contact of *G. duodenalis* with cells of the host's intestinal mucosa, facilitating effective colonization in the intestine[132].

Finally, it could be suggested that their genotypic distribution may also be related to the ability of the parasite genome to express variant surface proteins (VSPs), the main antigens recognized by the host, located on the surface of trophozoites. This antigenic variation is regulated by a mechanism of RNA interference (RNAi), which is still little known.[127]

Finally, this study provides greater knowledge about the genotypic distribution of the *G. duodenalis* parasite in humans in Brazil, raising questions about the influence of parasite-host interaction and the epidemiology of genotypes, as well as the possible zoonotic potential of the disease. The data obtained could increase information on the genotype frequency of this enteropathogen for future comparisons within different Brazilian regions and also in other countries. In addition, they provide clues for understanding, preventing and controlling the disease in question. Further studies of the molecular characterization of the *G. duodenalis* species in other states of the country are needed in order to gain a better understanding of the transmission dynamics and epidemiology of giardiasis.

5. BIBLIOGRAPHICAL REFERENCES

1. Appelbee AJ, Thompson RCA, Olson ME. *Giardia* and *Cryptosporidium* in mammalian wildlife - current status and future needs.Trends in Parasitol 2005; 21: 370-6.

2. Thompson RCA, Hopkins RM, Homan WL. Nomenclature and genetic groupings of *Giardia* infecting mammals. Parasitol Today 2000; 16: 210-3.

3. Sogayar MTL, Guimaraes S. *Giardia lamblia.* In: Neves, DP, Melo AL, Linardi PM, Vitor RWA. Human parasitology. 11.ed. Sao Paulo: Atheneu; 2005. p.121-6.

4. Levine ND, Corliss JO, Cox FG, Deroux G, Grain J, Honigberg BM, et al. A newly revised classification of the Protozoa J Protozool 1980; 27:37-58.

5. Adam RD. Biology of *Giardia lamblia*. Clinical Microbiol. Reviews 2001 14:447-75.

6. Kulda J, Nohynkova E. Flagellates of the human intestine and of intestines of other species. In: Kreier JP . Parasitic Protozoa. New York, Academic Press; 1978. p.2-139.

7. Andrews RH, Adams M, Boreham PF, Mayrhofer G, Meloni PB. *Giardia intestinalis*: electrophoretic evidence for a species complex. Int J Parasitol 1989; 19:183-190.

8. Meyer EA. Giardiasis: human parasitic diseases. Elsevier Science Amsterdam 1990; 32(Pt 2):368-72

9. Thompson RCA, Reynoldson J A, Mendis A H W. *Giardia* and giardiasis. Adv in Parasitol 1993; 32:71-160.

10. Marquardt WC, Dameree RS, Grieve RB. *Trichostrongyles* of cattle, sheep, goats, and other ruminants. In: Parasitology and Vector Biology. 2 ed. USA; 2000. p.122-4

11. Sousa MC, Morais JB, Machado JE, Poiares-da-Silva J. Genotyping of *Giardia lamblia* human isolates from Portugal by PCR-RFLP and sequencing. J Eukaryot Microbiol 2006; 53:174-6.

12. Franco RMBG, Cordeiro NS. Giardiasis and cryptosporidiosis in day-care centers in the municipality of Campinas, SP. Rev Soc Bras Med Trop 1996; 29:585-91.

13. Thompson RCA. The zoonotic significance and molecular epidemiology of *Giardia* and giardiasis. Vet Parasitol 2004; 126:15-35.

14. Marti M, Schraner EM, Wild P, Muller N, Knopf LG, Hehl AB. The secretory apparatus of an ancient eukaryote: protein sorting to separate export pathways occurs before formation of transient golgi-like compartments. Mol Biol Cell 2003a; 14:1433-47.

15. Simpson AG, Roger AJ, Silberman JD, Leipe DD, Edgcomb VP, Jermiin LS, et al. Evolutionary history of "early-diverging" eukaryotes: the excavate taxon Carpediemonas is a close relative of *Giardia*. Mol Biol 2002; 19:1782- 91.

16. Rey L. Parasitologiae doenças parasitarias do homem nos tropicos ocidental. 4.ed. Rio de Janeiro Guanabara Koogan 2008; p. 417-21.

17. Hausen MA, Freitas JC & Monteiro-Leal LH . The effects of metronidazole and furazolidone during *Giardia* differentiation into cysts. Exp Parasitol 2006; 135-41.

18. Cimerman B & Cimerman S. Parasitologia humana e seus fundamentos gérais. 2. ed. Sao Paulo. Atheneu 2005; p.28-33.

19. Schmidt GD, Roberts LS. Foundations of parasitology. 2ed. St. Louis, The C V Mosby Company, 1981.

20. Thompson RCA, Lymbery AJ, Meloni BP. Genetic variation in *Giardia* Kunstler, 1882 taxonomic and epidemiological significance. Protozool Abst 1990; 4:1-28.

21. Faculty of Pharmacy and Dentistry of Ribeirao Preto - USP. Available at URL: [accessed 20 Jun 2010]

http://www.fcfrp.usp.br/dactb/Parasitologia/Arquivos/Genero Giardia files/image007.jpg

22. Erlandsen, SL. Investigation into the life cycle of *Giardia* using videomicroscopy and field Emission SEM. In: Olson BE, Olson M E, Wallis PM. *Giardia*: The cosmopolitan parasite. 1ed. Wallingford, UK: 2002 CABI Publishing; 2002.

23. Benchimol, M. Participation of the adhesive disc during karyokinesis in *Giardia lamblia*. Biology of the Cell 2004; 291-301

24. World Health Organization. Parasitic zoonoses. Report of a Who Expert Committee with the Participation of FAO. Technical Report Series. 637. Geneva, 1979.

25. Jakubowski W, Graun GF. Update on the control of *Giardia* in water supplies. In: Olson B E, Olson M E, Wallis P M. *Giardia*: The cosmopolitan parasite.1ed. Wallingford, UK: 2002 CABI Publishing; 2002.

26. Heitman TL, Frederick LM, Viste JR, Guselle NJ, Morgan UM, Thompson RC. Prevalence of *Giardia* and *Cryptosporidium* and characterization of *Cryptosporidium spp.* isolated from wildlife, human, and agricultural sources in the North Saskatchewan River Basin in Alberta, Canada. Can J Microbiol 2002; 48:530-41.

27. Aquino LJ, Ferreira AW, Avila SL. In: Diagnòstico laboratorial das principais doenças

infecciosa e auto-immunes. 2 ed. Rio de Janeiro: Guanabara Koogan 2001; p. 250-54.

28. Wolfe MS. Giardiasis. Clin Microbiol Rev 1992; 5:93-100.

29. Gennari SM, Souza S. Giardiase. Sao Paulo Fort Dodge Saùde Animal 2002; p.13.

30. Sogayar R. Giardiasis. In: Ferreira AW, Avila SLM. Laboratory diagnosis of the main infectious and autoimmune diseases. 2.ed. Rio de Janeiro. Guanabara Koogan 2001; p.250-4.

31. Homan W, Mank T. Human giardiasis: genotype linked differences in clinical symptomatology. Int J Parasitol 2001; 31: 822-6.

32. Haque R, Roy S, Kabir M, Stroup SE, Mondal D, Houpt ER. *Giardia* assemblage A infection and diarrhea in Bangladesh. J Infect Dis 2005; 192: 2171-3.

33. Rossit AR, Almeida MT, Nogueira CA, Costa Oliveira JG, Barbosa DM, Moscardini AC, Mascarenhas JD, Gabbay YB, Marques FR, Cardoso LV, Cavasini CE, Machado RL. Bacterial, yeast, parasitic, and viral enteropathogens in HIV-infected children from Sao Paulo State, Southeastern Brazil. Diagn Microbiol Infect Dis 2007; 57:59-66.

34. Gonçalves AC, Gabbay YB, Mascarenhas JD, Yassaka MB, Moran LC, Fraga VD, Castro E, Franco C, Machado RL, Rossit AR. Calicivirus and *Giardia lamblia* are associated with diarrhea in human immunodeficiency virus-seropositive patients from southeast Brazil. Am J Trop Med Hyg. 2009; 81:463-6.

35. Hanevik K, Hausken T, Morken MH, Strand EA, MOrch K, Coll P, Helgeland L, Langeland N. Persisting symptoms and duodenal inflammation related to *Giardia duodenalis* infection. J Infect 2007;

36. Ankarklev J, Jerlstrom-Hultqvist J, Ringqvist E, Troell K, Svard SG. Behind the smile: cell biology and disease mechanisms of *Giardia* species. Nat Rev Microbiol 2010; Jun 8:413-22.

37. Shant J, Ghosh S, Bhattacharyya S, Ganguly Nk, Majumdar S. The alteration in signal transduction parameters induced by the excretory- secretory product from *Giardia lamblia*. Parasitol 2004; 129:421-30.

38. Federal University of Rio de Janeiro. Available at URL: [accessed 10 Jun 2010] http://acd.ufrj.br/labhac/figura12.htm

39. Federal University of Rio Grande do Sul. Available at URL: [accessed 10 Jun 2010] http://www6.ufrgs.br

40. Guerreiro NM, Herrera PM, Escalona L, Kolster CE, Yanes VG, Febres O, Naveda O, Naveda M. *Giardia lamblia*: comparison of two diagnostic methods and evaluation of response to treatment with metronidazole. G E N 1991; 45:105-10.

41. Dias LCS, Teixeira ATLS, Andrade M, Mendes CR, Palma JSR. Intestinal parasites: comparison between Coprotest and other stool examination methods. J Bras Patol 1998; 34: 239-41.

42. Wolfe MS. Clinical symptoms and diagnosis by traditional methods. In: Meyer E A. Giardiasis. Amsterdam; Elsevier 1990; 3:175-85.

43. Boghdadi AM, Sehly Ael D, Tadros MB, Mansour NS. Use and limitations of the enterotest in the diagnosis of *Giardia lamblia*. J Egypt Soc Parasitol 1988; 18:273-8.

44. Soliman MM, Taghi-Kilani R, Abou-Shady AF, El-Mageid SA, Handousa AA, Hegazi MM, Belosevic M. Comparison of serum antibody responses to *Giardia lamblia* of symptomatic and asymptomatic patients. Am J Trop Med Hyg 1998; 58:232-9.

45. Machado RL, Figueredo MC, Frade AF, Kudó ME, Silva Filho MG, Póvoa MM. Comparison of 4 techniques for the diagnosis of *Giardia lamblia* in stool of children from Belém city, Parà State, Brazil. Rev Soc Bras Med Trop 2001; 34:91-3.

46. Chappell CL, Matson CC. *Giardia* antigen detection in patients with chronic gastrointestinal disturbances. J Fam Pract 1992; 35:49-53.

47. Abbaszadegan M, Gerba CP, Rose JB. Detection of *Giardia* cysts with a cDNA probe and applications to water samples. Appl Environ Microbiol

48. Mahbubani MH, Bej AK, Perlin M, Schaefer FW 3RD, Jakubowski W, Atlas RM. Detection of *Giardia* cysts by using the polymerase chain reaction and distinguishing live from dead cysts. Appl Environ Microbiol 1991; 57: 3456-61.

49. Van Coppenraet B, Wallinga JA, Ruijs GJ, Bruins MJ, Verweij JJ. Parasitological diagnosis combining an internally controlled real-time PCR assay for the detection of four protozoa in stool samples with a testing algorithm for microscopy. Clin Microbiol Infect 2009; 15:869-74.

50. Marshall MM, Naumovitz D, Ortega Y.; Sterling CR Waterborne protozoan pathogens. Clin. Microbiol 1997; 10:67-85.

51. Cifuentes E, Gomez M, Blumenthal U, Tellez-Rojo MM, Romieu I, Ruiz-Palacios G, et al. Risk factors for *Giardia intestinalis* infection in agricultural villages practicing wastewater irrigation in Mexico. Am J Trop Med Hyg. 2000; 62:388-92.

52. WHO. The World Health Report. Fighting disease, fostering development. World Health Organization, Geneva, Switzerland. 1996.

53. Crompton DW. How much human helminthiasis is there in the world? J Parasitol. Review 1999; 85:397-403.

54. Karanis P, Kourenti C, Smith H. Waterborne transmission of protozoan parasites: A worldwide review of outbreaks and lessons learned. J Water Health. 2007; 5:1-38.

55. Thompson JD, Higgins DG, Gibson TJ. CLUSTAL W: improving the sensitivity of progressive multiple sequence alignment through sequence weighting, position-specific gap penalties and weight matrix choice. Nucleic Acids Res 1994; 22:4673-80.

56. CDC. Division of Parasitic Diseases Program Review. National Center for Infectious Diseases, Centers for Disease Control, Atlanta, GA, USA, S.C. 1995.

57. Ferson MJ. Infection and control in child care settings. Commun Dis Intell 1997; 21:333-7.

58. Thompson RCA, Robertson ID. Gastrointestinal parasites of dogs and cats: current issues. Compend Cont. Ed Prac Vet 2003; 25:4-11.

59. Johargy A, Ghazi H, Mumenah A. Frequency of viral, bacterial and parasitic enteropathogens among young children with acute diarrhoea in Saudi Arabia. J Pak Med Assoc 2010; 60:456-9.

60. Nùnez FA, López JL, de la Cruz AM, Finlay CM. Risk factors for *Giardia lamblia* infection in children in daycare centers in Havana, Cuba. Cad Saùde Pùblica 2003; 19:677-82.

61. Almeida AA, Delgado ML, Soares SC, Castro AO, Moreira MJ, Mendonça CM, Canada NB, et al. Genotype analysis of *Giardia* isolated from asymptomatic children in northern Portugal. J Eukaryot Microbiol 2006;53:177-8.

62. Karanis P, Opiela K, Al-Arousi M, Seitz HM. A comparison of phase contrast microscopy and an immunofluorescence test for the detection of *Giardia* spp. in faecal specimens from cattle and wild rodents. Trans Roy Soc Trop Med Hyg 1996; 90:250-1.

63. Prado MS, Strina A, Barreto ML, Oliveira-Assis AM, Paz LM, Cairncross S. Risk factors for infection with *Giardia duodenalis* in preschool children in the city of Salvador, Brazil. Epidemiol Infect 2003;131:899-906.

64. Schnack FJ, Fontana M, Barbosa PR, Silva LS, Baillargeon CM, Barichello T, Póvoa MM, et al. Enteropathogen associated with diarrheal disease in infants (< 5 years old) in a

population sample in greater metropolitan Criciùma, Santa Catarina State, Brazil. Cad Saùde Pùblica 2003; 19:1205-8.

65. Tashima NT, Simoes MJ. Enteroparasitic occurrence in fecal samples analyzed at the University of Western Sao Paulo-UNOESTE Clinical Laboratory, Presidente Prudente, Sao Paulo State, Brazil. Rev Inst Med Trop Sao Paulo 2004; 46:243-8.

66. Takizawa M G, Falavigna D L, Gomes M L. Enteroparasitosis and ethnographic relationship to food handlers in tourist and economic center in Paranà, Southern Brazil. Rev Inst Med Trop Sao Paulo 2009: 51:31-5.

67. Machado E R, Santos D S, Costa-Cruz J M. Enteroparasites and commensals among children in four peripheral districts of Uberlândia, state of Minas Gerais. Rev Soc Bras Med Trop 2008; 41:581-5.

68. Basso RM, Silva-Ribeiro RT, Soligo DS, Ribacki SI, Callegari- Jacques SM, Zoppas BC. Evolution of the prevalence of intestinal parasitosis among school children in Caxias do Sul, RS. Rev Soc Bras Med Trop 2008; 41:263-8.

69. Pittner E, Moraes IF, Sanches HF, Trincaus MR, Raimondo ML, Monteiro MC. Enteroparasitosis in children from a school community in the city of Guarapuava, PR. Revista Sallus 2007; 1:97-100.

70. Pereira MG, Atwill ER, Barbosa AP. Prevalence and associated risk factors for *Giardia lamblia* infection among children hospitalized for diarrhea in Goiânia, Goiàs State, Brazil. Rev Inst Med Trop Sao Paulo 2007; 49:139-45.

71. Carvalho-Costa FA, Gonçalves AQ, Lassance SL, de Albuquerque CP, Leite JP, Bóia MN. Detection of *Cryptosporidium* spp and other intestinal parasites in children with acute diarrhea and severe dehydration in Rio de Janeiro. Rev Soc Bras Med Trop 2007: 40:346-8.

72. Volotao AC, Costa-Macedo LM, Haddad FSM, Brandao A, Peralta JM, Fernandes O. Genotyping of *Giardia duodenalis* from human and animal samples from Brazil using beta-giardin gene : a phylogenetic analysis. acta Trop 2007; 102:258-62.

73. Silva RR, da Silva CAM, Pereira CAJ, Nicolato RLC, Negrao-Cornsa D, Lamounier J A, et al. Association between nutritional status, environmental and socio-economic factors and *Giardia lamblia* infections among children aged 6-71 months in Brazil. Trans Soc Trop Med Hyg. 2009; 103:512-9.

74. Tashima NT, Simoes MJ, Leite CQ, Fluminhan A, Nogueira MA, from a Malaspina AC.

Classic and molecular study of *Giardia duodenalis* in children daycare center in the region of Presidente Prudente, Sao Paulo, Brazil. Rev Inst Med Trop Sao Paulo 2009; 5:1924.

75. Machado et al. Giardiasis and Helminthiasis in children from day-care centers and schools (public and private) in the city of Mirassol, SP, Brazil. Rev Soc Bras Med Trop 1999 32:697-704.

76. Mascarini LM, Donalisio MR. Giardiasis and cryptosporidiosis in children institutionalized at daycare centers in the state of Sao Paulo. Rev Soc Bra Med Trop. 2006; 39:577-9.

77. Guimaraes S, Sogayar MI. Occurrence of *Giardia lamblia* in children of municipal day-care centers from Botucatu, Sao Paulo State, Brazil. Rev Inst Med Trop Sao Paulo 1995; 37:501-6.

78. Moura EC, Bragazza LM, Coelho MF, Aun SM. Prevalence of intestinal parasitosis in schoolchildren. J Pediatr 1997; 73:406-10.

79. Ludwig KM, Frei F, Alvares Filho F, Ribeiro-Paes JT. Correlation between sanitation conditions and intestinal parasitosis in the population of Assis, State of Sao Paulo. Rev Soc Bras Med Trop. 1999; 32:547-55.

80. Büyükbaba Boral O, Uysal H, Alan S & NAZLICAN O. Investigation of intestinal parasites in AIDS patients. Mikrobiyol Bul 2004; 38: 121-8.

81. Hung C C, Deng H Y, Hsiao W H. Invasive amebiasis as an emerging parasitic disease in patients with human immunodeficiency virus type infection in Taiwan. Arch intern Med 2005; 165: 409-15.

82. Ramakrishnan K, Shenbagarathai R, Uma, A. Prevalence of intestinal parasitic infestation in HIV/AIDS patients with diarrhea in Madurai City, South India. Jap J infect Dis 2007; 60: 209-10.

83. Cimerman S, Cimerman B & Lew DS. Prevalence of intestinal parasitic infections in patients with acquired immunodeficiency syndrome in Brazil. Int J infect Dis 1999; 3: 203-6.

84. Cardoso LV, Marques FR, Cavasini CE. Correlation of intestinal parasitic pathogens in HIV-seropositive adults with and without diarrhea in Northeast region of Sao Paulo State, Brazil. Rev panamer Infectol 2004; 6: 8-11.

85. Monis PT, Thompson RC. Cryptosporidium and *Giardia-zoonoses*: fact or fiction? Infect Genet Evol 2003; 3:233-44.

86. Thompson RC, Monis PT. Variation in *Giardia*: implications for taxonomy and

epidemiology. Adv Parasitol 2004; 58: 69-137.

87. Cacciò SM, Ryan U. Molecular epidemiology of giardiasis. Molecular & Biochem Parasitol 2008; 160:75-80.

88. Cacciò SM, Sprong H. *Giardia duodenalis*: genetic recombination and its implications for taxonomy and molecular epidemiology. Exp Parasitol 2010; 124:107-12.

89. Lalle M, Pozio E, Capelli G, Bruschi F, Crotti D, Cacciò SM. Genetic heterogeneity at the beta-giardin locus among human and animal isolates of *Giardia duodenalis* and identification of potentially zoonotic subgenotypes. Inter J Parasitol 2005b; 35: 207-13.

90. Sprong H, Cacciò SM, van der Giessen JW. ZOOPNET network and partners. Identification of zoonotic genotypes of *Giardia duodenalis*. PLoS Negl Trop Dis 2009; 3:e558.

91. Lalle M, Frangipane di Regalbono A, Poppi L, Nobili G, Tonanzi D, Pozio E, Cacciò SM. A novel *Giardia duodenalis* genotype A subgenotype in fallow deer. J Parasitol 2007; 93:426-8.

92. Thompson RC, Palmer CS, O'Handley R. The public health and clinical significance of *Giardia* and *Cryptosporidium* in domestic animals. Vet J 2008; 177:18-25.

93. Monis PT, Caccio SM, Thompson RCA. Variation in *Giardia*: towards a taxonomic revision of the genus. Trends Parasitol. 2009; 25:93-100.

94. Souza SLP, Gennari SM, Richtzenhain LJ, Pena HFJ, Funada MR Cortez A, Gregori F, Soares RM. Molecular identification of *Giardia duodenalis* isolates from humans, dogs, cats and cattle from the state of Sao Paulo, Brazil, by sequence analysis of fragments of glutamate dehydrogenase (gdh) coding gene. Vet Parasitol 2007; 149:258-64.

95. Minvielle MC, Molina NB, Polverino D, Basualdo JA. First genotyping of *Giardia lamblia* from human and animal feces in Argentina, South America. Mem Inst Oswaldo Cruz 2008; 103:98-103.

96. Abdel-Moneim SM, Sultan DM. Genetic characterization of *Giardia lamblia* isolates from Egyptian patients with relation to clinical giardiasis. J Egypt Soc Parasitol 2008; 38: 547-60.

97. Lalle M, Jimenez-Cardosa E, Cacciò SM, Pozio E. Genotyping of *Giardia duodenalis* from humans and dogs from Mexico using a betagiardin nested polymerase chain reaction assay. J Parasitol 2005a; 91:203-5.

98. Solarczyk P, Majewska AC. A survey of the prevalence and genotypes of *Giardia*

duodenalis infecting household and sheltered dogs. Parasitol 2010; 106:1015-19.

99. Paulina, R. C. Molecular detection of *Giardia* in fecal samples and water: extraction of genomic DNA, PCR and RFLP. 107 f. Thesis (Doctorate in Human and Animal Health) - Federal University of Paranâ. Technology Sector, Curitiba, 2005.

100. Amar CF, Dear PH, Pedraza-Diaz S, Looker N, Linnane E, Lauchlin J. Sensitive PCR-restriction fragment length polymorphism assay for detection and genotyping of *Giardia duodenalis* in human feces. J Clin Microbiol 2002;

40:446-52.

101. Bertrand I, Albertini L, Schwartzbrod J. Comparison of two target genes for detection and genotyping of *Giardia lamblia* in human feces by PCR and PCR- restriction fragment length polymorphism. J Clin Microbiol. 2005; 43:5940-4.

102. Sousa MC, Morais JB, Machado JE, Poiares-da-Silva J. Genotyping of *Giardia lamblia* human isolates from Portugal by PCR-RFLP and sequencing. J Eukaryot Microbiol 2006; 53:174-6.

103. Hopkins RM, Constantine CC, Groth DA, Wetherall JD, Reynoldson JA, Thompson RC. PCR-based DNA fingerprinting of Giardia duodenalis isolates using the intergenic rDNA spacer. Parasitol 1999; 118:531-9.

104. Cacciò SM, De Giacomo M, Aulicino FA, Pozio E. *Giardia* cysts in wastewater treatment plants in Italy. Appl Environ Microbiol 2003; 69:3393-8.

105. Traub RJ, Monis PT, Robertson I, Irwin P, Mencke N, Thompson RC. Epidemiological and molecular evidence supports the zoonotic transmission of *Giardia* among humans and dogs living in the same community. Parasitol 2004; 128:253-62.

106. Volotao AC, Souza Junior JC, Grassini C, Peralta JM, Fernandes O. Genotyping of *Giardia* duodenalis from Southern Brown Howler Monkeys (*Alouatta clamitans*) from Brazil. Vet Parasitol 2008; 158: 133-7.

107. Bertrand I, Schwartzbrod J. Detection and genotyping of *Giardia duodenalis* in wastewater: relation between assemblages and fecal contamination origin. Wat. Res 2007; 16:3675-82.

108. Graczyk TK, Thompson RC, Fayer R, Adams P, Morgan UM, Lewis EJ. *Giardia duodenalis* cysts of genotype A recovered from clams in the Chesapeake Bay subestuary Rhode River. Am J Trop Med Hyg 1999; 61:526-9.

109. Robertson LJ, Forberg T, Hermansen L, Gjerde BK, Langeland N. Molecular

characterization of *Giardia* isolates from clinical infections following a waterborne outbreak. J Infect 2007;55:79-88.

110. Plutzer J, Karanis P, Domokos K, Torokné A, Màrialigeti K. Detection and Characterization of *Giardia* and *Cryptosporidium* in Hungarian raw, surface and sewage water samples by IFT, PCR and sequence analysis of the SSU rRNA and GDH genes. Int J Hyg Environ Health 2008; 211:524-33.

111. Read C, Walters J, Robertson ID, Thompson RC. Correlation between genotype of *Giardia duodenalis* and diarrhoea. Int J Parasitol 2002; 32:229-31.

112. Yang R., Jeremy NG, Ryan U. High prevalence *Giardia duodenalis* assemblage B and potentially zoonotic subtypes in sporadic human cases in Western Australia. Int J of Parasitol 2009; 40:293-7.

113. Sahagùn J, Clavel A, Goni P, Serai C, Llorente MT, Castillo FJ, et al. Correlation between the presence of symptoms and the *Giarda duodenalis* genotype. Eur J Clin Microbiol Infect Dis 2007; 27:81-83.

114. Brazilian Institute of Geography and Statistics. URL: [accessed 10 Jul 2010] (http://www.ibge.gov.br/cidadesat/topwindow.htm?1).

115. Faust EC, Sawitz W, Tobie J, Odom V, Peres C, Lincicome DR. Comparative efficiency of various technics for the diagnosis of protozoa and helminth in feces. J Parasit 1939; 25: 241-62.

116. Keister BC. Axenic cultivation of *Giardia lamblia* in TYI-S-33 medium supplemented with bile. Trans R Soc Trop Med Hyg 1983; 77:487-8.

117. Jin L, Nei M. Limitations of evolutionary parsimony method of phylogenetic analysis. Mol Biol Evol 1990; 7:82-102.

118. Saitou N, Nei M. The neighbor-joining method: a new method for reconstructing phylogenetic trees. Mol Biol Evol 1997; 4:406-25.

119. Cacciò S, De Giacomo M, Pozio E. Sequence analysis of the *β-giardin* gene and development of a polymerase chain reaction-restriction fragment length polymorphism assay to genotype *Giardia duodenalis* cysts from human fecal samples. Intern J Parasitol 2002; 32:1023-30.

120. Slifko TR. Smith HV, Rose JB, et al. Emerging parasite zoonoses associated with water and food. Int JParasitol 2000; 30:1379-93.

121. Ankarklev J, Jerlstrom-Hultqvist J, Ringqvist E, Troell K. Behind the smile: cell biology and disease mechanisms of *Giardia* Species. Nat Rev Microbiol 2010; 10:1038-2317.

122. Ponce-Macotela M, Martinez G, Bermudez-Cru RM, Salazar-Schettino PM, EY PL. Unusual prevalence of the *Giardia intestinalis* A-II subtype among isolates from humans and domestic animals in Mexico. J Parasitol 2002; 120112.

123. Page N, Pager C, Steele AD. Characterization of rotavirus strains detected in Windhoek, Namibia during 1998-1999. J Infect Dis 202; 28:162-7.

124. Chirdan OO, Akosu JT, Adah SO. Intestinal parasites in children attending day care centers in Jos, Central Nigeria. Niger J Med. 2010 ;192:219-22.

125. Ajjampur SSR, Kannan A, Sathyakumar K, Sarkar R, Gladstone BP, Kang G. *Giardia duodenalis* Assemblages Associated with Diarrhea in Children in South India Identified by PCR-RFLP. Am J Trop Med Hyg 2009; 80:16-19.

126. Nash T. E., Herrington DA, Losonsky GA, Levine MM. Experimental human infections with *Giardia lamblia*. J. Infect. Dis 1987; 156: 974-84.

127. Prucca CG, Lujan HD. Antigenic variation in *Giardia lamblia.* Cel Micro 2009; 11:1706-15.

128. Barbara JD, Housley MP, Smith JR, Andersen YS, Martin GM. Polymeric immunoglobulin receptor in intestinal immune defense against the lmen- dwelling protozoan parasite *Giardia*. J Immun 2010; 177:6281-90.

129. Kamda JD, Singer SM. Phosphoinositide 3-Kinase-Dependent Inhibition of Dendritic Cell Interleukin-12 Production by *Giardia lamblia.* Infect Immun 2009; 77:685-93.

130. Sousa MC, Gonçalves CA, Bairos VA, Poiares-Da-Silva J. Adherence of *Giardia lamblia* trophozoites to Int-407 human intestinal cells. Clin Diagn Lab Immunol 2001; 8:258-65.

131. Troeger H, Epple HJ, Schneider T, Wahnschaffe U, Ullrich R, Burchard GD, et al. Effect of chronic *Giardia lamblia* infection on epithelial transport and barrier function in human duodenum. Gut 2007; 56:328-35.

132. Ringqvist E, Palm JE, Skarin H, Hehl AB, Weiland M, Davids BJ, et al. Release of metabolic enzymes by *Giardia* in response to interaction with intestinal epithelial cells. Mol Biochem Parasitol 2008; 1592:85-91.

MIX
Papier aus verantwortungsvollen Quellen
Paper from responsible sources
FSC® C105338

Printed by Books on Demand GmbH, Norderstedt / Germany